Of Caffeine, Acetaminophen and Immunomodulator Adenosine

Rowena Kong

2020

First Printing: 2020

ISBN: 9798640563085

Table of Contents

Chapter 1: Neuroprotective Functions: The Potential Roles of Caffeine and Acetaminophen in Delayed Progression of Parkinson's Disease

They are too common in our everyday life and language to be ignored. When physical symptoms call for our action, we reach for the bottle of Tylenol, whereas to satisfy our craving, we aim for the kitchen coffee-brewer or the nearest cafe. They are *effective* and have always been so. But are these remedial actions just about all or the best that acetaminophen(the active ingredient in Tylenol) and caffeine can offer us? It seems that scientific research is beginning to paint the fuller picture. Experts are tapping into the drugs' less obvious but potential useful functions that could be relevant in the treatment of Parkinson's disease, a neurodegenerative disorder which debilitates an individual's motor and muscular ability. While the causes of Parkinson's disease remain unknown, genetic factors do not appear to contribute to a majority of the cases and epidemiologic studies have also focused on studying health and lifestyle factors and their correlations with the risk of disease development (Carlson, 2011; Simola, Pinna, Frau, & Morelli, 2014).

Studies which reported that participants' caffeine intake lowered one's risk of developing Parkinson's disease led to promising proposed theories of mechanisms of how adenosine receptor A_{2A} antagonists provide neuroprotective benefits by modulating the excitotoxicity of glutamate-induced activity in the subthalamic nucleus (Simola et al., 2014). They also indirectly attenuate neuroinflammation through action on microglia and astrocytes, cells which provide protective and supportive functions to neurons. It has been known that caffeine renders us more alert by blocking adenosine receptors, A_1 and A_{2A}, as the level of adenosine increases with periods of wakefulness and helps promote sleepiness (Carlson, 2011; Kong et al., 2002). In the case of Parkinson's disease, the focus is on the multiple negative effects of stimulated A_{2A} receptors on nigrostriatal dopaminergic system, the pathway involved in movement control, which connects the substantia nigra with the striatum and a target region of treatment against the disease. The degeneration of dopaminergic neurons in this region contributes to the motor deficit symptoms of Parkinson's disease and is thought to increase glutamatergic input from the cortex to the striatum. This in turn elevates glutamatergic activity in the subthalamic nucleus and results in excitotoxicity effect on neurons in the substantia nigra, which are very close by in the region. Studies have demonstrated that the death of dopaminergic neurons in the subthalamic nucleus and decrease in dopamine levels in the striatum can be counteracted by A_{2A} antagonists as the stimulation of such receptors increased the level of extracellular glutamate (Greenamyre, 2001; Lancelot & Beal, 1998; Morelli et al., 2010; Popoli, Betto, Reggio, & Ricciarello, 1995; Schwarzschild et al., 2003). From this proposed theory, we get to have a glimpse of the intimate interconnection between different brain regions which function together to affect our neurophysiological well-being.

As mentioned, second mechanism of A_{2A} antagonists works by countering neuroinflammation through suppressing the activation of microglia and astrocytes, which also express A_{2A} receptors and initiate inflammatory responses (Armentero et al., 2011; Halliday and Stevens, 2011; Hirsch and Hunot, 2009; Litteljohn, Mangano, Clarke, Bobyn, Moloney, & Hayley, 2010; Lopes, Sebastião, & Ribeiro, 2011; Reale et al., 2009). Neuroinflammation plays a role in the neurodegeneration progress of Parkinson's disease. It is worth a note that the action of of A_1 receptor antagonists did not produce comparable results of neuroprotective benefits as those of A_{2A} receptors in mice models of Parkinson's disease (Chen et al., 2001).

As multiple processes are involved in the progression of Parkinson's disease, there are treatment options which may target certain mechanisms but at the same time, not impacting others. Therefore, a single drug or surgical procedure may not suffice to keep the full range of symptoms at bay. The degeneration of dopaminergic neurons, apart from its initiation by genetic mutation and protein misfolding and aggregation which are followed by a series of cellular disarray that culminates in motor dysfunction of an individual, is also exacerbated by oxidative stress. Research has looked into the potential role of the over-the-counter pain remedy, acetaminophen in reducing the effect of such process on disease progression. In an animal model study, administration of low concentrations of acetaminophen have been shown to be protective against neurodegeneration induced by 6-hydroxydopamine(6-OHDA), a neurotoxin which can self-oxidize and generate reactive oxygen species to deplete anti-oxidant enzymes within cells and this leads to eventual cell damage (Locke, Fox, Caldwell, & Caldwell, 2008; Simola, Morelli, & Carta, 2007). In addition, various concentrations of acetaminophen in the same study were effective in suppressing tyrosine hydroxylase(TH)-induced degeneration, also a contributor to oxygen radical formation that harms dopaminergic neurons (Adams Jr., 2012; Locke et al., 2008).

A study by Tripathy and Grammas (2009) which tested the response of rat brain endothelial cells, that were pre-treated with acetaminophen, to oxidative stress demonstrated increased cellular survival when exposed to menadione, the stressor which released reactive oxygen species. The protection offered to these cells was also due in part to the ability of acetaminophen to increase the expression of an anti-apoptotic protein Bcl2 and thus negatively influenced cell death. A later study reported that the addition of acetaminophen to the brain microvessels of rats increased vascular expression of neuroprotective proteins, which were discovered to decrease with increasing age of the animals (Tripathy, Sanchez, Yin, Martinez, & Grammas, 2012). Although further research is needed, and not just animal models but also human epidemiologic studies of acetaminophen consumption and in spite of the drug's mechanisms being less direct and target-specific compared with caffeine or A_{2A} antagonists, such preliminary evidence should not be taken lightly just because they sound trivial and too good to be true.

Chapter 2: Caffeine and Memory Consolidation

We all know coffee, with its caffeine content, that it is an effective stimulant which increases our arousal and keeps us awake while we pull our all nighter during grueling finals (and midterms for tougher courses). It is 'addictively useful' for both exams and socializing, regardless of whether one is a university student or not, and not surprisingly, researchers are also baffled by its popularity that they could not help but attempt to dig deeper under its superficial appeal in search of possible hidden benefits for our health. And it was not without success, for progress has been made in reports of caffeine's associations with neuroprotective effects against Parkinson's disease, Alzheimer's disease, and even potentially beneficial against chronic and fatty liver diseases. These benefits are still under ongoing study, and while we do appreciate the long-term value of such discoveries, as students, we are likely more concerned about how coffee can impact us academic-wise.

It turns out that caffeine research in the area of cognition and attention is not a new thing, though results obtained have been mixed. Depending on the condition under which learning takes place, caffeine can be a mild cognitive enhancer due to its cumulative influence on arousal, task performance and concentration, among other factors. Still, that does not deter students from consuming it. I, for one, do so on a regular basis, for the lame excuse of mood improvement. However, I have also found another reason for not quitting, and that is after reading a recent research report of how caffeine may even help promote memory consolidation, a process which converts information encoded in short-term memory into long-term storage (Borota et al., 2014; Kelly, Mikell, & McKhann, 2014). A group of researchers at Johns Hopkins and University of California, Irvine, led by Dr.

Michael Yassa, tested study participants by administering a surprise memory test a day after they were shown a set of over a hundred pictures and given a pill which contained either caffeine or a placebo. Unlike some previous studies, participants took the pill after they were acquainted with the pictures, for the purpose of testing the effect on memory consolidation and the results did show that participants who consumed moderate doses of caffeine were better able to discriminate between the first and a new set of pictures shown the day after. In other words, these participants were able to tell more accurately than those who received the placebo that certain items in the new picture set were only 'similar' instead of 'old' or have already appeared in the previous set. Somehow, this helps shed a positive light on the potential memory-enhancing effects of caffeine.

While research has looked into the positive effects of caffeine on working memory, people may still be skeptical of its unclear and likely mechanisms, though most focused on neurotransmitters adenosine(involved in wakefulness) and dopamine(involved in pleasure and reward learning) receptors, and how it can be generalized to more complex long-term memory beyond simple object recognition tasks. There is also the implication of a role of dopamine in memory processing which is emphasised less in the literature, since caffeine tends to act indirectly on dopamine receptors through its antagonism on adenosine receptors. But just in case you are wondering if you should be drinking coffee during or after regular, unrushed quiet study times rather than only during highly stressful final weeks(even though caffeine has been shown to lessen the negative effects of stressors and sleep deprivation and improve cognitive performance at the same time), I hope the above research could provide you with enough incentive to experiment and experience the personal results for yourself. After all, every student should be developing healthy study habits when aiming for success, whether she has a fondness for coffee or not.

Chapter 3: Adenosine and its Energy-Conserving Strategy

Adenosine is known as a neuromodulator with a broad range of functions across neurophysiological systems in cerebral, vascular, renal, immunological and muscular health, to name a few. Intracellular and extracellular adenosine can be generated by several sources and pathways, but more commonly through phosphohydrolysis of the high energy stores of extracellular adenosine triphosphate and adenosine diphosphate (ATP/ADP) by enzymes CD39 (nucleoside triphosphate dephosphorylase) and CD73 (ecto-5'-nucleotidase). There is ample research documenting possible neuroprotective mechanisms performed by adenosine through its receptor activation that sets into motion cascading processes which aid in promoting and maintaining neural and immune health (Alam, Costales,

& Cavanaugh et al., 2015; Bauerle et al., 2011,; Deaglio et al., 2007). However, lesser known and discussed is the implicit energy store regulating role performed by adenosine through various channels and receptor transmission that can be both simply understood and complex. Based on the close connection of adenosine with its origin precursors of the family of adenosine phosphates (ATP/ADP/Adenosine monophosphate[AMP]), it could be proposed a fundamental strategy of energy conservation mechanism by adenosine through a process of interactive recycling and replenishing that persists in maintaining a consistent storehouse of ATP-sourced energy which caters for a host of balanced neurophysiological functions. In a sense, adenosine is a significant and crucial player as a relaying mid-point between its precursors and reconsolidated phosphate variant products such that the continuous chain of energy supply can be maintained and regulated in an efficient and uninterrupted manner under normal homeostatic surveillance of neurophysiological conditions.

In this discussion, an energy conservation strategy proposed to apply to the relevant neurochemical elements, is an implicit mechanism that helps preserve a regulated flow of energy production maintained through various efficiently orchestrated neurochemical reactions that source, exchange and reform precursor and product molecular units with the goal of minimising excess energy expenditure and maximising reusability. Under such proposed definition, an example can be demonstrated by the involvement of adenosine in dampening the process of inflammation through dephosphorylation of ATP, which increases during tissue inflammation, by enzymes CD39 and CD73 (Alam, Costales, & Cavanaugh et al., 2015;

Deaglio et al., 2007). Such mechanism implies a pathway of depletion of extracellular ATP which is consequently being converted to adenosine as the end product to discontinue the process of inflammation that could potentially further tax cellular energy stores and harm the human host of a pathogenic invasion. It can be inferred that ATP is a necessary resource supply for the performance of the immune response task of inflammation and that the halting of such work by its conversion to adenosine therefore helps set the brake on the process, be it of immediate necessity or leading to a prospective long-term significance based on an individual's physiological capacity to provide and endure, thereby strengthening the concept of energy-conserving role played by adenosine. Nevertheless, such strategy does not preclude the reformation of ATP from adenosine at other potential sites of need after a temporal gap in order to maintain a steady flow of the energy pool through various channels.

In conclusion, the human physiological system is an integrative and autonomous entity that is almost entirely energy resource-dependent based on consideration of the needs and demands for sustenance of multiple make-ups of such a system that works in a complex fashion of intensive and orchestrated cooperation. Through such careful perspective and analysis of energy provision and needs equation, it is hoped that scientific endeavours that aim to better our understanding of human physiology could build more robust and sound component principles that lead to promising and evidence-based results.

Diagram 1: Breakdown of Action of Neuro-element on Energy Pool

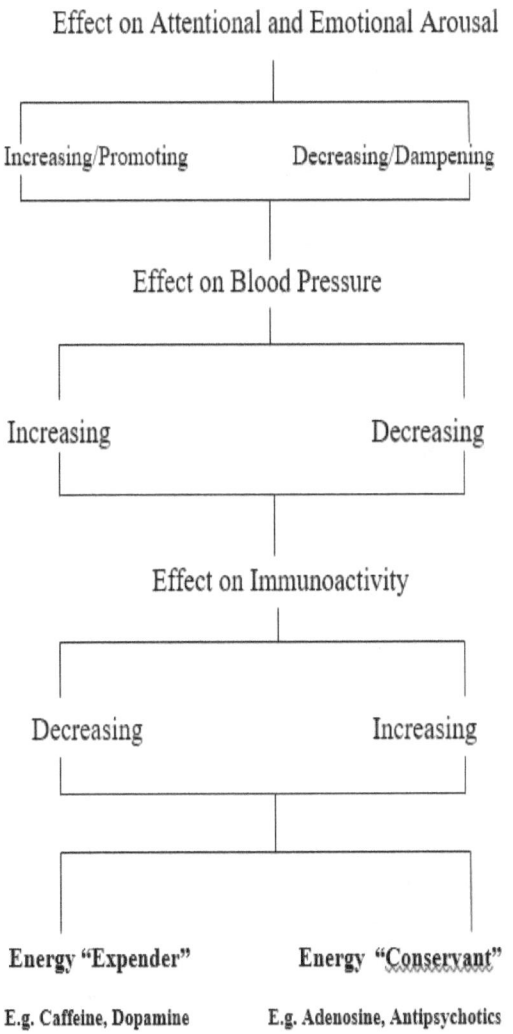

Effect on Attentional and Emotional Arousal

Increasing/Promoting   Decreasing/Dampening

Effect on Blood Pressure

Increasing   Decreasing

Effect on Immunoactivity

Decreasing   Increasing

Energy "Expender"   Energy "Conservant"

E.g. Caffeine, Dopamine   E.g. Adenosine, Antipsychotics

References

Adams Jr., J. D. (2012). Parkinson's disease - apoptosis and dopamine oxidation. *Open Journal of Apoptosis, 1(1),* 1-8. doi:10.4236/ojapo.2012.11001

Alam, M. S., Costales, M. G., Cavanaugh, C., & Williams, K. (2015). Extracellular adenosine generation in the regulation of pro-inflammatory responses and pathogen colonization. *Biomolecules, 5*(2), 775-792.

Armentero, M. T., Pinna, A., Ferré, S., Lanciego, J. L., Müller, C. E., & Franco, R. (2011). Past, present and future of A(2A) adenosine receptor antagonists in the therapy of Parkinson's disease. *Pharmacology and Therapeutics, 132,* 280–299. doi:10.1016/j.pharmthera.2011.07.004

Bauerle, J. D., Grenz, A., Kim, J. H., Lee, H. T., & Eltzschig, H. K. (2011). Adenosine generation and signaling during acute kidney injury. *Journal of the American Society of Nephrology, 22*(1), 14-20.

Borota, D., Murray, E., Keceli, G., Chang, A., Watabe, J. M., Ly, M., ... & Yassa, M. A. (2014). Post-study caffeine administration enhances memory consolidation in humans. Nature neuroscience, 17(2), 201-203.

Carlson, N. R. (2011). *Foundations of behavioral neuroscience.* Boston, MA: Allyn and Bacon.

Chen, J., Xu, K., Petzer, J. P., Staal, R., Xu, Y., Beilstein, M.,...Schwarzschild, M. A. (2001). Neuroprotection by caffeine and A2A adenosine receptor inactivation in a model of Parkinson's disease. *The Journal of Neuroscience, 21,* 1-6.

Deaglio, S., Dwyer, K. M., Gao, W., Friedman, D., Usheva, A., Erat, A., ... & Kuchroo, V. K. (2007). Adenosine generation catalyzed by CD39 and CD73 expressed on regulatory T cells mediates immune suppression. *Journal of Experimental Medicine, 204*(6), 1257-1265.

Greenamyre, J. T. (2001). Glutamatergic influences on the basal ganglia. *Clinical Neuropharmacology, 24,* 65–70.

Halliday, G. M., & Stevens, C. H. (2011). Glia: Initiators and progressors of pathology in Parkinson's disease. *Movement Disorders, 26,* 6–17. doi: 10.1002/mds.23455

Hirsch, E. C., & Hunot, S. (2009). Neuroinflammation in Parkinson's disease: A target for neuroprotection? *Lancet Neurology, 8,* 382–397.

Kelly, K. M., Mikell, C. B., & McKhann, G. M. (2014). Morning Joe or After-Dinner Espresso? Improved Memory Consolidation After Caffeine Administration. Neurosurgery, 74(6), N8-N11.

Kong, J., Shepel, N., Holden, C. P., Mackiewicz, M., Pack, A. I., & Geiger, J. D. (2002). Brain glycogen decreases with increased periods of wakefulness: Implications for homeostatic drive to sleep. *Journal of Neuroscience, 22,* 5581-5587.

Lancelot, E., & Beal, M. F. (1998). Glutamate toxicity in chronic neurodegenerative disease. *Progress in Brain Research, 116,* 331–347.

Litteljohn, D., Mangano, E., Clarke, M., Bobyn, J., Moloney, K., & Hayley, S. (2010). Inflammatory mechanisms of neurodegeneration in toxin-based models of Parkinson's disease. *Parkinson's Disease, 2011,* 1-18. doi:10.4061/2011/713517

Locke, C. J., Fox, S. A., Caldwell, G. A., & Caldwell, K. A. (2008). Acetaminophen attenuates dopamine neuron degeneration in animal models of Parkinson's disease. *Neuroscience Letters, 2,* 129-133. doi:10.1016/j.neulet.2008.05.003

Lopes, L. V., Sebastião, A. M., & Ribeiro, J. A. (2011). Adenosine and related drugs in brain diseases: Present and future in clinical trials. *Current Topics in Medicinal Chemistry, 11,* 1087–1101.

Morelli, M., Carta, A. R., Kachroo, A., & Schwarzschild, M. A. (2010). Pathophysiological roles for purines: Adenosine, caffeine and urate. *Progress in Brain Research, 183,* 183–208. doi:10.1016/S0079-6123(10)83010-9

Popoli, P., Betto, P., Reggio, R., & Ricciarello, G. (1995). Adenosine A(2A) receptor stimulation enhances striatal extracellular glutamate levels in rats. *European Journal of Pharmacology, 287,* 215–217.

Reale, M., Iarlori, C., Thomas, A., Gambi, D., Perfetti, B., Di Nicola, M., & Onofrj, M. (2009). Peripheral cytokines profile in Parkinson's disease. *Brain, Behavior, and Immunity, 23,* 55–63. doi:10.1016/j.bbi.2008.07.003

Schwarzschild, M. A., Chen, J. F., Tennis, M., Messing, S., Kamp, C., Ascherio, A., Holloway, R. G., Marek, K., Tanner, C. M., McDermott, M., Lang, A. E., & The

Parkinson Study Group. (2003). Relating caffeine consumption to Parkinson's disease progression and dyskinesias development. *Movement Disorders, 18,* 1082–1083. doi:10.1002/mds.10585

Simola, N., Morelli, M., & Carta, A. R. (2007). The 6-Hydroxydopamine model of Parkinson's disease. *Neurotoxicity Research, 11,* 151-167.

Simola, N., Pinna, A., Frau, L., & Morelli, M. (2014). Protective agents in Parkinson's disease: Caffeine and adenosine A2A receptor antagonists. In R. M. Kostrzewa (Ed.), *Handbook of Neurotoxocity* (pp. 2281-2298). New York, NY: Springer

Tripathy, D., & Grammas, P. (2009). Acetaminophen protects brain endothelial cells against oxidative stress. *Microvascular Research, 77,* 289-296. doi:10.1016/j.mvr.2009.02.002

Tripathy, D., Sanchez, A., Yin, X., Martinez, J., & Grammas, P. (2012). Age-related decrease in cerebrovascular-derived neuroprotective proteins: Effect of acetaminophen. *Microvascular Research, 84,* 278-285. doi:10.1016/j.mvr.2012.08.004

www.ingramcontent.com/pod-product-compliance
Lightning Source LLC
Chambersburg PA
CBHW030602220526
45463CB00007B/3142